Apples:
A Social History

SALLY TWISS

The National Trust

FOR PHOEBE, THE APPLE OF MY EYE

First published in Great Britain in 1999 by
National Trust Enterprises Ltd
36 Queen Anne's Gate, London SW1H 9AS
http://bookshelf.nationaltrust.org.uk

British Library Cataloguing in Publication Data
A catalogue record for this book is available from the British Library.
ISBN 0 7078 0340 3
Designed and typeset by Peter and Alison Guy
Production by Dee Maple
Print managed by Centurion Press Ltd (HGP)

Front Cover: A display of apples from the Walled Garden at Berrington Hall, Herefordshire.
Clockwise from top left: Crimson Queening, Ashmead's Kernel, Catshead, Downton Pippin.
Title Page: *A Child with an Apple* by Jean-Baptiste Greuze,
on display at Upton House, Warwickshire.
Back Cover: *Cider-making at Westcott* by Mary Martin on display at Cotehele in Cornwall.

The Magic Apple Tree *by Samuel Palmer.*

Introduction

My interest in apples dates back to my childhood. Our large Somerset garden was planted with a rich variety of apples and the strawberry-like taste of that first bite into a juicy sun-warmed Worcester Pearmain, and the sight of the Devonshire Quarrenden laden with its small dark crimson apples remains with me. I can still remember the exhilaration at harvest time of climbing to the topmost branches of the Sheep's Nose cooker and Egremont Russet to reach the best fruit – always just a fingertip away – to throw down to the catchers below.

Apples seem to inspire such intense memories in many people, or to spark new passions often bordering on the obsessive. And who can fail to be intrigued by the wonderful names of old apple varieties or seduced by the luscious descriptions of their flavour and appearance? The crimson-gold freckles and stripes of Lady Sudeley, fragrant aroma of Cox's Pomona, names like Cornish Gilliflower, Peasgood Nonsuch and King's Acre Bountiful. This book attempts to capture the diversity of the apple and I hope it will tempt the uninitiated to savour its fascination for themselves.

Thanks to National Trust staff who have been so helpful and whose interest and knowledge is inspiring, including Barry Champion, Ted Bullock, Nick Winney, Lewis Eynon, Mark Pethullis and Paul Sharman whose notes on the Killerton cider orchards were invaluable; to Helen Fewster for the picture research and encouragement and to Phillip and Susie for their support.

The age-old story of Adam and Eve in the Garden of Eden is depicted in this detail of a cabinet at Sudbury Hall, Derbyshire. The serpent is shown coiled around the apple tree. Historians now believe that the forbidden fruit was more likely to have been a fig or a pomegranate, rather than an apple.

The apple is more firmly rooted in our history and culture than any other fruit. References to apples date back to the Old Testament; although the fruit which tempted Eve is more likely to have been a fig or pomegranate, it is in Deuteronomy that the expression 'the apple of his eye' is first coined, while the Song of Solomon asks 'Comfort me with apples, for I am sick of love'. More precisely, the apple is thought to have originated in south-west Asia and was first cultivated as far back as 1500 BC. During the first millennium BC, in a foretaste of more modern times, epicures in Persia would taste and compare the flavour and attributes of the different varieties available in those times, and no feast was complete without apples and other fruit to savour and admire.

Cultivated apples were introduced to Britain by the Romans, who found the wild apples growing here too sour. The Romans had developed skills in propagating trees through grafting and budding – techniques still used today – and during the invasion, soldiers were given land to plant orchards and for other uses as an inducement to stay. These orchards were abandoned as the Roman Empire fell and Britain was overrun by succeeding hordes of Angles, Saxons and Jutes. But during the Dark Ages orchards survived in the monasteries, which were self-sufficient in apples. Another invasion, this time by the Normans, saw new impetus in the cultivation of apples in Britain. The Normans had a long tradition of apple growing and cider making and brought their techniques with them. New varieties of apples were introduced to Britain; including the Costard, from which the word Costermonger is derived, originally meaning a seller of Costard apples, and the Pearmain – records show that in 1290 the manor of Runham, Norfolk, paid the Exchequer 200 Pearmains and 4 hogsheads of cider made from Pearmains.

The Victorians celebrated the apple for its diversity of flavour and decorative qualities. The nineteenth-century writer, artist and designer William Morris used Pomona, the Roman goddess of fruits and seeds, as the inspiration for one of his tapestry designs, displayed at Wightwick Manor in the West Midlands.

Apple-growing underwent another expansion during the reign of Henry VIII. His fruiterer Richard Harris brought new varieties such as pippins over from France and set up the first commercial orchard, which supplied trees to wealthy landowners. Fruit growing became fashionable and well established, often under the influence of succeeding monarchs; John Tradescant, gardener to Charles I, travelled as far as Russia to bring back new varieties and planted over a thousand fruit trees in a garden created for the king's French wife, Henrietta Maria. Apples were used in all manner of dishes and the trees were planted as decorative and productive elements in pleasure grounds.

Towards the end of the eighteenth century, poor orchard management and changes in agriculture brought about a decline in apple production: dairying and cereal growing were more profitable. However, during Victorian times, interest in particular varieties revived and the apple once again became the supreme fruit. Bon viveurs considered it an epicurean delight, comparing the aroma and flavour of different apples with as much gravitas as they would consider a fine wine. Old and valued varieties were tracked down and propagated and new ones were nurtured; Cox's Orange Pippin and Bramley's Seedling, mainstays of the modern apple industry, were developed during Victorian times. A more scientific approach to apple-growing saw the establishment of the fruit research station at Long Ashton, Somerset, and later, a second site at East Malling in Kent.

The twentieth century has seen a major decline in the orchards of Britain. Changes in agricultural practice, development for housing, industry and roads have all contributed to this decline; according to Ministry of Agriculture statistics, around 150,000 acres (60,750 ha) – two-thirds of the total – of commercial orchards have been destroyed since 1960. Devon, one of the major apple-growing areas, has lost almost ninety per cent of its orchards in the last twenty-five years. The demands of the apple industry dictate that it is only viable to grow and sell a few varieties and until recently the rich heritage of apples was under threat.

A collection of around fifty apple varieties are grown in the Walled Garden at Berrington Hall in Herefordshire, including Court of Wick (foreground) and Doctor Hare's.

Today the popularity of the apple and appreciation of its diversity is once again in the ascendancy. The environmental and arts charity Common Ground has played a major part through its Save Our Orchards campaign and instigation of Apple Day, which has alerted consumers to the fact that there are many other varieties besides Cox's Orange Pippin and Bramley's Seedling. Although it is unlikely that we will ever see large-scale commercial orchard planting of standard trees again, old orchards are being restored and replanted by individuals and communities, sometimes aided by local authorities which have set up schemes to safeguard and conserve orchards and help towards the cost of planting regional varieties of fruit. One of the most innovative is in the South Hams, between the southern slopes of Dartmoor and the sea. Through their Environment Service, the District Council launched a programme in 1989 to save the remaining orchards in this part of Devon by providing trees from their nursery of local varieties, giving training in pruning and planting, and organising events to raise awareness of the disappearing heritage of apples like Totnes Apple and Doll's Eye. The project has been taken one step further with the setting up of collective marketing and cider-making.

Over 2,400 varieties of apples, together with hundreds of cultivars of cherries, plums, pears and other fruit are now safe with the Brogdale Horticultural Trust in Kent which is renowned throughout the world for its work in the conservation of fruit. In addition to breeding and trialing new varieties for the commercial fruit industry, Brogdale encourages amateurs to appreciate and enjoy apples through a comprehensive range of events. Formerly the Brogdale Experimental Horticulture Station, it was threatened by closure until the Horticultural Trust, an independent, non-profit making organisation, took it over from the Ministry of Agriculture in 1990.

At that time, when closure looked possible, the National Council for the Conservation of Plants and Gardens (NCCPG) started a survey of fruit varieties held in small collections and by amateur gardeners with the aim of saving them from extinction. It succeeded in the case of the Herefordshire variety, Tillington Court, a brightly coloured cooking apple. Graftwood was taken from the last known tree of its kind, grown on and the resulting two trees planted in the Walled Garden at the National Trust's Berrington Hall in Herefordshire, where the NCCPG has put together a collection of local varieties for their historical interest and as a gene pool for the future. Most of the varieties represented at Berrington were introduced before the turn of the century and some date back several hundreds of years.

The orchard in bloom is a beautiful sight and an important feature of the landscape, as illustrated by this picture of The Courts in Wiltshire. It can also be a haven for wildlife like bees, butterflies and birds.

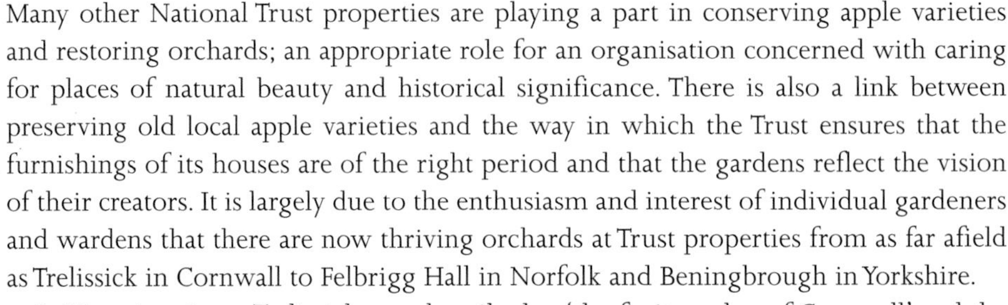

Many other National Trust properties are playing a part in conserving apple varieties and restoring orchards; an appropriate role for an organisation concerned with caring for places of natural beauty and historical significance. There is also a link between preserving old local apple varieties and the way in which the Trust ensures that the furnishings of its houses are of the right period and that the gardens reflect the vision of their creators. It is largely due to the enthusiasm and interest of individual gardeners and wardens that there are now thriving orchards at Trust properties from as far afield as Trelissick in Cornwall to Felbrigg Hall in Norfolk and Beningbrough in Yorkshire.

In Victorian times, Trelissick was described as 'the fruit garden of Cornwall' and the Head Gardener Barry Champion ensures that it lives up to its name. He is building up a gene bank of old Cornish and Tamar Valley apples and has travelled hundreds of miles to take grafts of varieties such as Cornish Gilliflower, Mannacan Primrose and Polly Whitehair. The orchard has been planted in a quincunx pattern, where lines of trees radiate away from the eye wherever one stands, and an avenue of apples is also being planted. No chemicals are used on the trees and the orchard is primarily for wildlife – birds and butterflies feast on the fallen fruit in the autumn, and bees visit the apple blossom and Cornish daffodils growing beneath in the spring.

The eighteenth-century Walled Garden at Felbrigg is well-established with both standard and wall-trained apples, most of which are East Anglian culinary varieties such as Norfolk Beefing, Golden Noble and Norfolk Royal Russet, a modern sport of the older Norfolk Royal. The soil at Felbrigg is very light and the Head Gardener Ted Bullock and his team have to incorporate plenty of organic matter and water the wall trees frequently in their first year or two. Following a long-held tradition at Felbrigg, these are trained as fans, rather than espaliers where the branches are trained horizontally. There is a fine display of espaliered fruit at Westbury Court in Gloucestershire using varieties which date back to the seventeenth century, when the decorative effect of apples was highly valued. The west wall in autumn is a wonderful sight, with its rows of foliage and apples ranging in colour from the yellow Lemon Pippin to the crimson Devonshire Quarrenden, the russet Golden Reinette, huge green Catshead and tiny red flushed Api Rose.

In the north, a new orchard of Lake District varieties such as Forty Shilling, Longstart and Lemon Square – an apple peculiar to the Eden Valley – has been planted at Acorn Bank in Cumbria as a result of local interest following an Apple Day, while at Beningbrough and Nunnington, orchards are being rejuvenated or created and planted with apples from Yorkshire.

Killerton Sweet, which originated on the National Trust's Killerton estate in Devon, one of the cider producing counties of the West Country. Many other good varieties of cider, dessert and culinary apples come from Devon including Devonshire Quarrenden, Allspice, Plympton Pippin, Tremlett's Bitter and Fair Maid of Devon.

One of the most ambitious of the National Trust's projects to rejuvenate orchards is at Killerton in Devon where about four hundred fruit trees, mainly old local apple varieties, have been planted around the estate during the last ten years. At the turn of the century, over one hundred acres of the 7000-acre estate were planted to orchards in this rich, fertile area of Devon. These had gradually declined, and this traditional aspect of local agricultural history was in danger of being lost forever, but the great gale of January 1990, which so devastated areas of the West Country including the garden and many of the orchards at Killerton, provided the impetus for replanting and sparked the enthusiasm of wardens Paul Sharman and Bill Lambshead.

Their first task was to try and identify the remaining trees, and information provided by retired estate employees and local farmers proved invaluable. A veritable treasure trove of old and unusual varieties was uncovered, including the intriguingly named cider apples Slack-Ma-Girdle and Hangy Down. Even more exciting was the discovery that three of the varieties had originated on the estate – Killerton Sweet and Killerton Bitter, both cider apples, and the pretty little red dessert apple Star of Devon. Grafts were taken by a nurseryman who was also able to supply other local varieties on standard rootstocks. The old orchards at Killerton contained a mixture of apples, pears, plums, quinces and damsons and this tradition has continued, again using local varieties often grafted from existing old but declining trees on the estate. The success of the scheme has also seen a revival in cider-making here (see page 28) which is bottled and sold in the Killerton shop and served in all National Trust restaurants in Devon.

Killerton Cider Chutney
(makes about 4 lb/2 kg)
Highfield Preserves of Tiverton make vast quantities of chutney using the cider made on the estate. This is a scaled-down version of their recipe:

4lb (2kg) peeled and chopped apples
1½lb (675g) chopped onions
12oz (350g) preserving sugar
12oz (350g) raisins
juice of 2 lemons
½ pint (275ml) cider vinegar
½ pint (275ml) Killerton cider
pinch of chilli powder
1 teaspoon ground ginger
½ teaspoon salt

Cut the fruit into small pieces. Cook all the ingredients in a preserving pan or a large saucepan by bringing to the boil, then cooking slowly until thick. The chutney is ready when a spoon drawn through the mixture leaves a trail. Pot in sterilised jars, seal and cover. Keep in a cool dark cupboard.

For those who would rather buy than try, Killerton Cider Chutney is available at the National Trust shop at Killerton.

Onion Redstreak, illustrated by the artist Mary Martin, who discovered one of the last trees of this Tamar Valley variety when the apple was brought to an Apple Day at Cotehele.

Unlike the other south-western counties of Devon and Somerset, Cornwall has never been a major apple producing area. To some misguided experts from upcountry, the chance of growing apples successfully in the mild, damp climate, often poor soil and salt laden winds was nigh on impossible. And yet Cornwall has a diversity of excellent local apples; its once isolated position ensured that many of the varieties were specific to particular areas and that most farms and homes were self-sufficient in apples with at least a few trees, if not an orchard. Shelter belts were planted to protect the blossom from strong winds and the orchard was a haven for early spring lambs and bees.

Cornwall seems to have been unique in producing a variety of apples for pickling which was considered a local delicacy in the western part of the county. This tradition is almost lost, but the signs of a revival for Cornish orchards is good with many enthusiasts taking up the baton for local varieties, particularly in the Tamar Valley, which was once famous for its apple and cherry orchards. The artist Mary Martin and her partner James Evans have been instrumental in tracking down and propagating old varieties of apple, pear and cherry and act as advisers at Cotehele. Here the old dairy orchard is being replanted with local varieties including the Onion Redstreak, a near extinct Tamar apple which appeared at an Apple Day held at the property. Just below at Cotehele quay, where paddle steamers once called to take Tamar Valley fruit to Devonport market in Plymouth, a new orchard was created in 1995 to celebrate the National Trust's Centenary. The project fired the imagination of several local people who each paid for a tree (or trees) of such varieties as Hocking's Green, Cornish Pine and Collogett Pippin and helped to plant them in the Centenary Orchard.

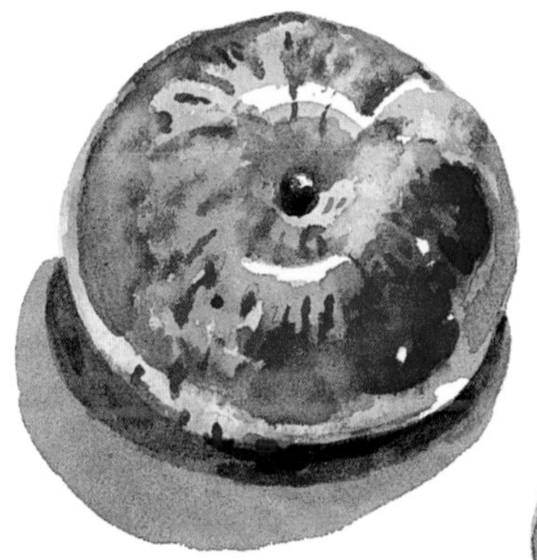

Onion Redstreak

Norfolk Beefing, or Biffin, growing at Berrington. This useful culinary variety will keep until spring, when it will be sweet enough to eat for dessert.

The dark red Norfolk Beefing (pronounced Biffin) is a cooking apple but, kept until the spring, will be sweet enough to eat as a dessert apple. The rather dry flesh was ideal for cutting into rings to be dried to prolong its use even further or for baking, as the fruit would not burst. When cooked very slowly for several hours (traditionally in a bread oven) the apple had a rich, almost spicy flavour and was served as a Christmas sweetmeat.

Norfolk Beefing has a long history; it was first recorded in Norfolk in 1807 but a 'beefing' is listed in a late seventeenth-century notebook of the Walpole family of Norwich. In his diary written at the end of the eighteenth century, Parson Woodforde frequently refers to his 'Beefans' apples, supplies of which would be sent to succour his parishioners. Although not as common in Norfolk as it once was, trees of this variety can still be found in the county, such as those growing in the Walled Garden at Felbrigg Hall. Norfolk Beefing is also grown at Berrington Hall where the fruit can be compared with the Herefordshire Beefing.

Another large East Anglian apple, Lady Henniker, was valued not only for its lovely flavour but for its appearance too. The bright yellow-green fruit was raised in the 1840s at Lord Henniker's estate, Thornham Hall in Suffolk, by his gardener John Perkins who used it 'when large and handsome dishes of mixed fruit are required. Its appearance by lamplight is most telling'. John Perkins carefully selected the tree as the most promising of a number of seedlings grown from pips in cider pomace (apple pulp). In 1873 his efforts were rewarded when it received a First Class Certificate from the Royal Horticultural Society. Lady Henniker also grows at Felbrigg, where the collection includes D'Arcy Spice, Golden Noble, Winter Majetin and Dr Harvey, named after a seventeenth-century Master of Trinity Hall, Cambridge.

An apple falling to the ground inspired Sir Isaac Newton to formulate the Theory of Gravity. The living remains of the tree, believed to be Flower of Kent, have recently been re-discovered in the grounds of Newton's family home at Woolsthorpe Manor in Lincolnshire.

Flower of Kent is thought to be the apple variety which inspired Sir Isaac Newton's Theory of Gravity. In 1665-6, Newton was staying at his mother's house, Woolsthorpe Manor in Lincolnshire, and whilst sitting in the garden watched an apple fall from a tree. It was probably this incident that started Newton wondering if the same force was also responsible for keeping the moon orbiting the earth and thus to the discovery of universal gravitation. The original tree fell victim to a storm in 1820 and it was believed that only cuttings survived, some at universities around the world. However, the recent discovery of a tree trunk in the exact spot, which had re-rooted at both ends, has led scientists to hope that Newton's tree lives on. Not only scientists either; Woolsthorpe Manor is now owned by the National Trust and the tree has become something of a shrine to Newton. The Flower of Kent apple which provoked such a significant step in the history of science is generally a large fruit and would certainly have dropped to earth with a hefty thud.

Many apples have interesting histories in the way that they were cultivated or brought to public notice, even if not with such far-reaching implications as those on Newton's tree. The handsome dessert apple Lady Sudeley was raised in 1849 at Petworth in Sussex by Mr Jacob, an agricultural worker, and was originally named Jacob's Strawberry after him. The beauty of the apple, with its bold red stripes of crimson over a greenish yellow-gold base, was said to have caught the eye of Lord Sudeley when he was having an extensive orchard of this variety planted in Gloucestershire. The pattern and colours of the fruit reminded him of a dress worn by his wife at Court and the apple was subsequently renamed.

Another version of this renaming attributes it to the nurseryman George Bunyard who, having introduced the apple, was supplying Lord Sudeley with the trees for his orchards and named it after his best customer. Whichever is true, the apple is certainly worth growing for its appearance and juicy good flavour but it must be eaten almost as soon as it is picked, otherwise it becomes dry with a slightly acid after-taste.

Ten Commandments is something of novelty variety which was raised in 1883 in Herefordshire. When cut in half from side to side, ten red spots spaced in a circle around the core can be seen; hence its name. It was used mainly for cider making.

Herefordshire has a wealth of wonderful varieties of apples. It was a group of local enthusiasts from Hereford who helped to start a nationwide programme of apple identification, which was to prove vital in combating the invasion of foreign imported apples in the mid-1870's. Members of the Woolhope Naturalists' Field Club organised a survey of orchards in the county to track down all varieties to find out which were the best; with this information recommendations could be made as to which were most suitable to cultivate in the locality and would be the most popular and profitable to grow to meet market demand. The project expanded to include apples and pears from all over the country and the Hereford Show became the place for enthusiasts to go, with hundreds of varieties being displayed.

Today a few of those varieties can be seen growing at Berrington Hall as part of its fruit collection. One of the most beautiful is Crimson Queening (see front cover) which was first recorded in 1831 but is probably much older. The aromatic fruit was highly esteemed for display in shows as well as for its usefulness in the kitchen. In Herefordshire it is still sometimes called Quoining because of its angular shape; an angle is known as a quoin in the building trade.

Two connoisseur apples to be found growing at Berrington are Pitmaston Pine Apple and Yellow Ingestrie. The Pine Apple is a delightful little amber fruit with a rich nutty taste although the famous pomologist Dr Robert Hogg noticed a distinct pineapple flavour. It was raised in 1785 by Mr White, the steward of Lord Foley of Stoke Edith in Herefordshire where there was to be a well known fruit collection in later years when another good dessert apple, Stoke Edith Pippin, was raised. Yellow Ingestrie is named after the colour of its skin and Ingestrie Hall in Staffordshire. It was much esteemed for its highly flavoured fruit, which has an almost wine-like taste, and for the decorative appearance of the apples and tree itself. It was raised by Thomas Andrew Knight of Downton Castle in Shropshire who also raised Downton Pippin (see front cover) and Wormsley Pippin, both of which grow at Berrington.

Catshead is named after its distinctive shape; when viewed from the side it resembles the profile of a cat.

Described by the poet John Philips (1679-1709) as a 'weighty orb' in his poem 'Cyder', Catshead is the apple for making dumplings. Its angular shape was just right for wrapping up in pastry to make a filling midday meal for farm workers in the nineteenth century. Huge old trees of Catshead can still be seen growing in the Midlands, from where it originates, as can those of Bess Pool, another local variety. This handsome apple, with its crimson mottled skin, is named after the girl who found it growing in a Nottinghamshire wood. Bess Pool was the daughter of an inn-keeper near Chilwell and her namesake went on to achieve local fame as a well-flavoured fruit for cooking, eating and display. The tree was introduced by a local nurseryman and was widely grown in the nineteenth century particularly in the Midlands and north, where its very late flowering period could escape the late frosts.

The popularity of Catshead, Bess Pool and other culinary apples has long been supplanted by another Nottinghamshire apple, Bramley's Seedling. Despite its ubiquity today, its origins were more the result of chance than of the type of scientific research which produces our modern apples. Bramley's Seedling was raised from a pip of unknown origin by a Miss Brailsford and planted in a cottage garden in Southwell about 190 years ago. Here it grew in relative obscurity until some years later, by which time the local butcher, Mr Bramley, owned the cottage. The nurseryman Henry Merryweather spotted the tree and subsequently introduced it. From such humble beginnings, the Bramley went on to be awarded a First Class Certificate from the Royal Horticultural Society in 1883 and to become almost the only culinary apple grown on a commercial basis in the United Kingdom and available throughout the year.

Apple Dumplings

Apple dumplings are quick and easy to make – another reason for their popularity as a farm worker's lunch. Pastry is wrapped around the peeled and cored apple, then tucked into the core hole and brushed with milk. The dish is baked in a fairly hot oven until the apple is soft. The apple can be stuffed with butter, sugar, jam, raisins or any number of fillings before it is parcelled up.

In Germany, apple dumplings are called 'apples in dressing gowns' (Apfel im Schlafrock). In *Grace Before Meat* (1821) Charles Lamb wrote that the poet Samuel Taylor Coleridge 'holds that a man cannot have a pure mind who refuses apple-dumplings'.

Ribston Pippin is a delicious dessert apple with a good balance of sugar and acid; properties which were inherited by its more famous offspring, the Cox's Orange Pippin. Ribston can be found in several National Trust orchards including those at Barrington Court in Somerset, Berrington Hall, Beningbrough and Coughton Court, Warwickshire.

Orchards are particularly associated with the West Country and south-eastern counties of England; apples are not grown commercially in the north. Domestic orchards were once a common feature of the landscape here, however, and many apples originated in areas such as Cheshire, Cumbria and over the border in Scotland including Lord Derby, Keswick Codlin and James Grieve. Indeed it was the pip from a Yorkshire apple, the Ribston Pippin, which produced our most widely grown dessert apple, the Cox's Orange Pippin. The Ribston Pippin was found growing at Ribston Hall, near Knaresborough, in the early eighteenth century and is believed to have been raised from a pip brought over from France. By the nineteenth century it was one of the most popular apples and was planted all over the country. The apple was prized for its juiciness and strong aromatic flavour; the nurseryman and epicure Edward Bunyard advised that, rather than eat the apple straight from the tree, the Ribston Pippin should be stored and tasted frequently in order to capture the moment when acid and sugar are in perfect balance. Its popularity gradually declined and was overtaken commercially by its famous offspring.

At Beningbrough Hall in Yorkshire, Ribston Pippin is trained as an espalier. Other north of England varieties growing here as espaliers, cordons or pyramids include Golden Spire, Cockpit, Carlisle Codling and Yorkshire Greening.

The Cox's Orange Pippin needs little description; its attraction lies in the intensity of flavour and keeping qualities. It was raised in 1825 by a retired brewer, Richard Cox, at Colnbrook Lawn, near Slough and won great acclaim, being voted best dessert apple from southern England in 1883. Unlike its parent, the Cox is not suitable for growing in the north and requires the warmer climate of south-eastern counties to crop well. Its susceptibility to scab, mildew and canker saw a decline in its fortunes at the beginning of the twentieth century until chemical sprays to combat these diseases were introduced in the 1920s.

Preventing Scurvy

Before citrus fruit were imported to this country, whaling ships would dock in Whitby to load up with good keeping apples from Yorkshire orchards to help prevent the sailors from contracting scurvy.

Tom Putt (below) and Hoary Morning (right) are handsome dual purpose apples which originate from Somerset.

The evocatively named Hoary Morning originated from Somerset and was first recorded in 1819. It is a handsome apple with red stripes over a yellowish ground and with a good flavour for cooking and eating as a dessert apple. It was also a good keeper, staying in good condition until late winter or early spring if stored carefully. With these attributes, it is no surprise that Hoary Morning was once a favourite West Country apple; sadly, it is now hard to find.

Another popular Somerset apple, Tom Putt, is much more readily available. It was raised in the late eighteenth century by the Reverend Thomas Putt, who was rector of Trent in Somerset and later at Farway, near Honiton, Devon. Tom Putt apple trees can be bought at some specialist nurseries and are still to be found in West Country orchards. At Killerton it is one of the varieties used to make cider. A number of trees have been planted on private estates in the vicinity of the Reverend Putt's final home in East Devon to maintain its presence in the locality.

It is certainly a tree worth planting if you have the space; well-shaped with attractive blossom and beautiful fruit, with one half streaked bright red and the other yellowish green. Although highly esteemed as a cider apple, it is equally good for cooking and as a juicy dessert apple. Tom Putt was the favourite choice of 'scrumping' school boys and this is probably one of the reasons why the apple was given several other names including Devonshire Nine Square, Ploughman, Izods Kernel and Marrow Bone.

Somerset was especially well-known for its cider apples and many of the best have originated in the county, including Dabinett, Somerset Redstreak and Yarlington Mill – so named because it grew out of a wall by the water wheel at Yarlington, near West Cadbury – and the most famous cider apple of all, the Kingston Black or Black Taunton, which produces a vintage cider of particular flavour.

The cider-making tradition continues at Killerton, as warden Paul Sharman demonstrates. An apple mill crushes the fruit, then a 'cheese' of alternate layers of pulp and hessian are built up and squeezed on the 150-year-old cider press to extract the juice (see detail). Straw would once have been used instead of hessian but modern straw proved to be too short for the press. The juice is left to ferment, then racked off and filtered every three months to get rid of 'snarlydogs' – the residue not already thrown out during the fermentation process. As one year's brew is pressed, the previous year's is being bottled.

Like the apple, the popularity of cider has waxed and waned over the centuries. It was first introduced to Britain by the Normans after the invasion in 1066. France has a long tradition of producing the drink – one of the first references to cider was made by Charlemagne at the beginning of the ninth century. It quickly became popular with all sections of society; most manor houses had their own presses and monasteries planted orchards to supply their needs and supplement their income: records from Battle Abbey in Sussex show that, in 1369, three tuns of cider were sold for 55 shillings.

Over the next three centuries, cider-makers had mixed fortunes. But in the mid-seventeenth century when agriculture went into a decline, cider was seen as an alternative source of revenue. Cider orchards were planted throughout the country, and production techniques were examined. The finest cider is produced by selecting particular varieties, and apples such as Foxwhelp and Redstreak were propagated and planted. Cider became the champagne of England and the national drink: the best rivalled foreign wines for quality; vintage cider was imbibed by the wealthy, and farmhouse cider by the rest of society. Farm workers were paid partly in cider, as well as other farm produce, and large quantities were drunk during the working day, especially at harvest time. This continued into the twentieth century, when the increasing mechanisation of agriculture made it dangerous and illegal.

The very popularity of the drink contributed to its decline in the eighteenth century, as merchants went for quantity rather than quality. At the same time, a tax on cider was introduced and together with new agricultural developments, cider-making gradually became less profitable for farmers. It saw a brief revival in Victorian times when a more scientific approach was developed and taken up by small new cider factories such as Bulmers in Herefordshire and Taunton Cider in Somerset. Since then, apart from the occasional spell when cider has become fashionable, production has fallen with orchards making way for arable crops and dairy farming after the Second World War and excise duty raised so that only specialist or large producers could survive. Even the latter are finding it difficult; two West Country manufacturers were recently bought out by much bigger concerns and subsequently closed down.

Specialist cider-makers are once more coming into their own; 'cider for sale' signs are still quite a common sight in the West Country although unsuspecting visitors are often surprised at the difference between the commercially bottled variety and the rougher stronger farm cider! There has also been a revival of distilled cider – originally produced in the seventeenth century – by Burrow Hill Cider in Somerset.

The apple lends itself to a huge diversity of recipes, perhaps most famously to apple pie. Different varieties will bring a different flavour to the dish; Bunyard advised that James Grieve would make a 'pie full of subtle memories'. In the nursery rhyme, here illustrated by Kate Greenaway, the letters of the alphabet fight over possession of the apple pie.

Since early times, cider and apples have been highly regarded for their health giving properties. Indeed, cider was credited as having restorative powers and imbibing regularly was thought to be the secret of a long life, according to a Devonshire drinking song:

> I were brought up on cider
> An I be a hundred and two
> But still that be nuthin' when you come to think
> Me father and mother be still in the pink
> An they were brought up on cider
> Of the rare old Tavistock brew
> An me granfer drinks quarts
> For he's one of the sports
> That were brought up on cider too

The saying 'An apple a day keeps the doctor away' is certainly based on fact and apples were used as cures for all manner of ills. In medieval times, apples provided relief from constipation and in the seventeenth century from coughs. The celebrated Elizabethan herbalist Gerard recommended its use for the complexion: ' there is an ointment made with the pulp of apples and swine's grease and rose water, which is used to beautify the face, and to take away the roughness of the skin, called in shops pomatum of the apples whereof it is made'.

The digestive power of the apple has long been recognised. It is a good source of dietary fibre, which is mostly contained in the skin. A medium sized Bramley provides about one-fifth of the daily recommended amount of fibre while a Cox's Orange Pippin gives one-tenth of the fibre needed in a daily diet. Apples are low in calories, consisting of eighty-four per cent water and some varieties are particularly high in vitamin C; the Ribston Pippin contains 30 milligrams of vitamin C for every 100g of fruit and the Sturmer Pippin 29 milligrams. Apples also contain flavonoid, an anti-oxidant which is believed to prevent the furring of the arteries.

There is every reason to eat an apple (or several) a day either fresh or in the seemingly endless variations of recipes based on the fruit. It is the most versatile of fruits, working well with meat – where would roast pork be without apple sauce? – cheese, vegetables (apples and red cabbage are a winter staple in northern parts of Europe) and salads, in pies, puddings and refreshing cordials. Using distinct varieties can add a whole new dimension of flavour to familiar dishes – try baking the culinary apple Dr Harvey, which becomes sweeter and highly flavoured, instead of the more acidic Bramley, and just taste the difference.

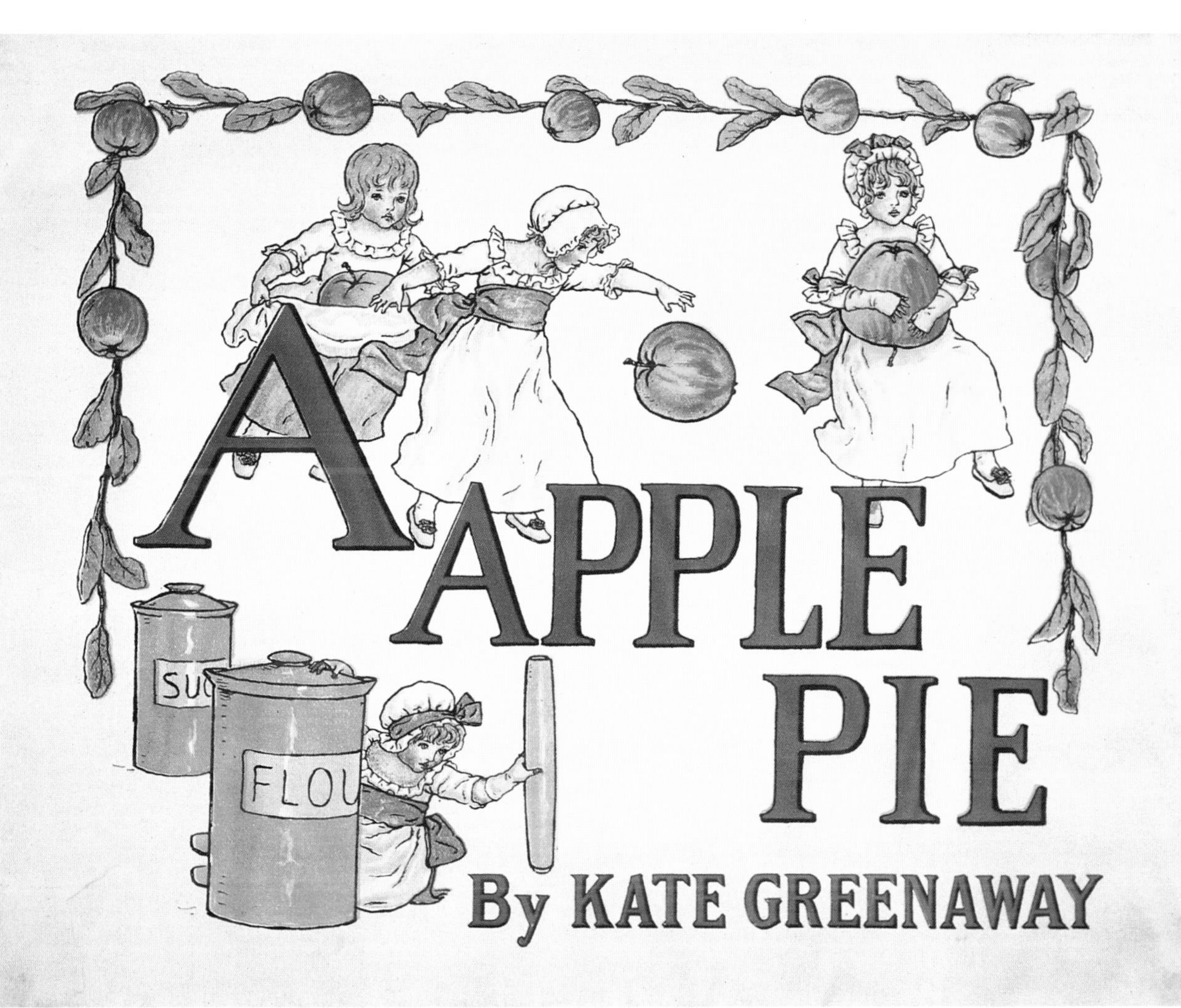
A
APPLE
PIE
By KATE GREENAWAY

A cartoon by Kate Charlesworth about Apple Day, which first appeared in the New Scientist *in 1993.*

One of the most significant events which has helped to rekindle interest in old apple varieties is the introduction of Apple Day. The environmental charity Common Ground launched a Save Our Orchards campaign in 1989 and initiated Apple Day the following year on 21st October. Its aim was to celebrate local apples and orchards and the diversity of landscape and culture linked to them. Since that first Apple Day, which was held in London's Covent Garden, the idea has been embraced by organisations and individuals throughout the country.

Enthusiasm for the event has been matched by inventiveness. The diversity of the apple has been celebrated through tastings, guided orchard walks and talks, cookery and pruning demonstrations, cider-making and apple bobbing, photographic exhibitions and recipe exchanges. One of the added bonuses of Apple Day has been re-discovering local varieties that were thought to be lost, such as Red Rollo and White Quarantine which were taken for identification at an Apple Day in Cornwall. Even fictional apples have got in on the act; when BBC Radio's serial 'The Archers' featured an Apple Day, one of the characters thought he had found the rare Borsetshire Beauty in his orchard.

Apple Day is now an established date on the calendar, but there are many much older traditions associated with apples. On New Year's Day in the Forest of Dean it was customary to give 'applegifts' as tokens of friendship and good health. The apples were stuck with oats, wheat or raisins and decorated with nuts and sprigs of evergreen such as box or yew, and mounted on three sticks to form a stand. St Thomas's Day, 21 December, was another date when apples were a traditional gift, as recalled in the song: 'Bud well, bear well/God send farewell/A bushel of apples to give/On St Thomas's morning'. Apples are closely linked with Halloween; it is said that if you eat an apple then, you will not get a cold for twelve months. But it is Twelfth Night which is particularly significant in the apple calendar, for this is when the wassail takes place.

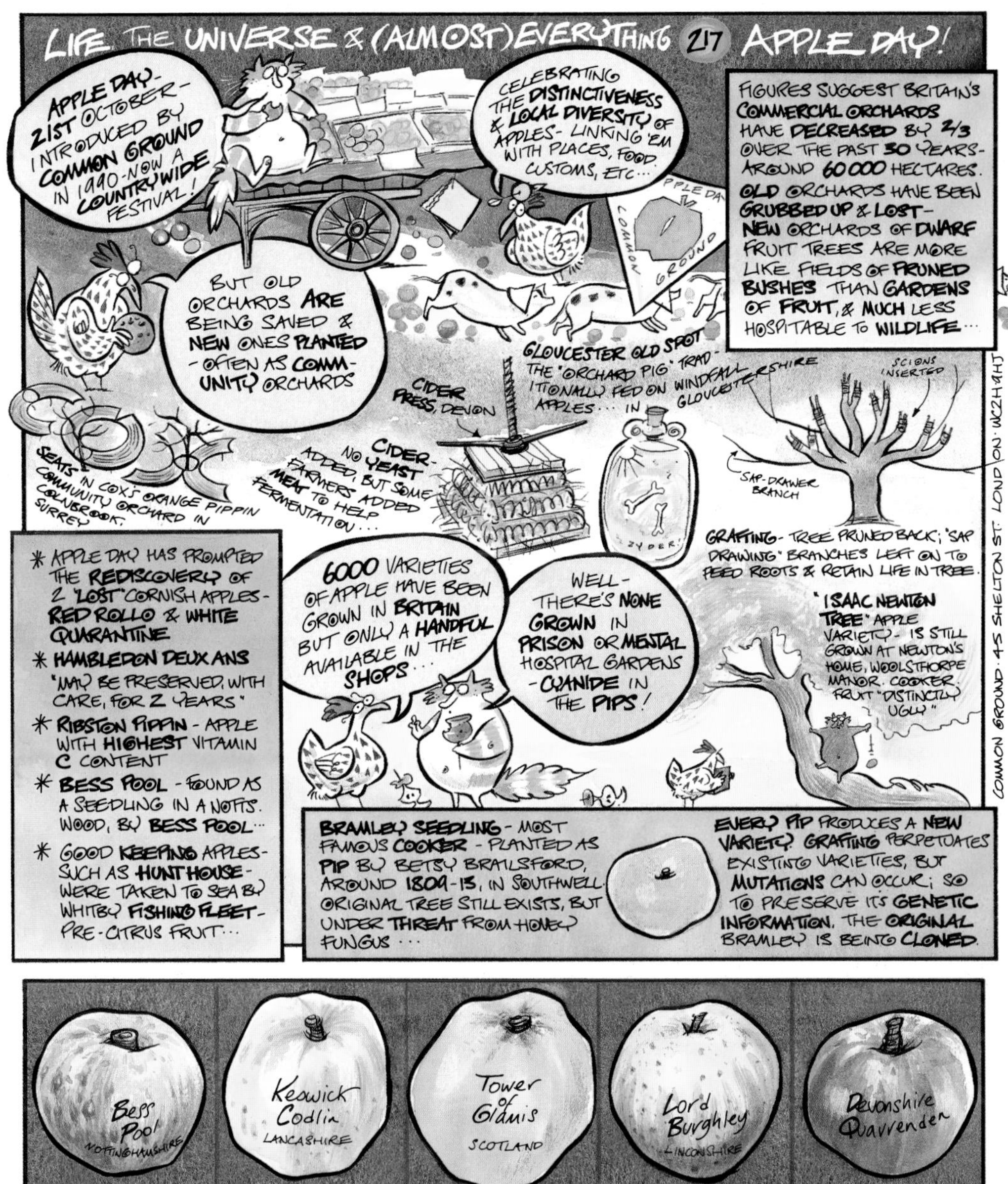
LIFE, THE UNIVERSE & (ALMOST) EVERYTHING 217 APPLE DAY!
APPLE DAY - 21ST OCTOBER - INTRODUCED BY COMMON GROUND IN 1990 - NOW A COUNTRYWIDE FESTIVAL!
CELEBRATING THE DISTINCTIVENESS & LOCAL DIVERSITY OF APPLES - LINKING 'EM WITH PLACES, FOOD CUSTOMS, ETC...
APPLE DAY COMMON GROUND
FIGURES SUGGEST BRITAIN'S COMMERCIAL ORCHARDS HAVE DECREASED BY 2/3 OVER THE PAST 30 YEARS - AROUND 60 000 HECTARES. OLD ORCHARDS HAVE BEEN GRUBBED UP & LOST - NEW ORCHARDS OF DWARF FRUIT TREES ARE MORE LIKE FIELDS OF PRUNED BUSHES THAN GARDENS OF FRUIT, & MUCH LESS HOSPITABLE TO WILDLIFE...
BUT OLD ORCHARDS ARE BEING SAVED & NEW ONES PLANTED - OFTEN AS COMMUNITY ORCHARDS
GLOUCESTER OLD SPOT - THE "ORCHARD PIG" TRADITIONALLY FED ON WINDFALL APPLES... IN GLOUCESTERSHIRE
CIDER PRESS, DEVON
SCIONS INSERTED
SAP-DRAWER BRANCH
SEATS IN COX'S ORANGE PIPPIN COMMUNITY ORCHARD IN COLNBROOK, SURREY
CIDER - NO YEAST ADDED, BUT SOME FARMERS ADDED MEAT TO HELP FERMENTATION...
ZYDER!
GRAFTING - TREE PRUNED BACK; "SAP DRAWING" BRANCHES LEFT ON TO FEED ROOTS & RETAIN LIFE IN TREE.
COMMON GROUND · 45 SHELTON ST. LONDON · WC2H 9HJ
* APPLE DAY HAS PROMPTED THE REDISCOVERY OF 2 "LOST" CORNISH APPLES - RED ROLLO & WHITE QUARANTINE
* HAMBLEDON DEUX ANS "MAY BE PRESERVED, WITH CARE, FOR 2 YEARS"
* RIBSTON PIPPIN - APPLE WITH HIGHEST VITAMIN C CONTENT
* BESS POOL - FOUND AS A SEEDLING IN A NOTTS. WOOD, BY BESS POOL...
* GOOD KEEPING APPLES - SUCH AS HUNT HOUSE - WERE TAKEN TO SEA BY WHITBY FISHING FLEET - PRE-CITRUS FRUIT...
6000 VARIETIES OF APPLE HAVE BEEN GROWN IN BRITAIN BUT ONLY A HANDFUL AVAILABLE IN THE SHOPS...
WELL - THERE'S NONE GROWN IN PRISON OR MENTAL HOSPITAL GARDENS - CYANIDE IN THE PIPS!
"ISAAC NEWTON TREE" APPLE VARIETY - IS STILL GROWN AT NEWTON'S HOME, WOOLSTHORPE MANOR. COOKER. FRUIT "DISTINCTLY UGLY"
BRAMLEY SEEDLING - MOST FAMOUS COOKER - PLANTED AS PIP BY BETSY BRAILSFORD, AROUND 1809-13, IN SOUTHWELL. ORIGINAL TREE STILL EXISTS, BUT UNDER THREAT FROM HONEY FUNGUS...
EVERY PIP PRODUCES A NEW VARIETY. GRAFTING PERPETUATES EXISTING VARIETIES, BUT MUTATIONS CAN OCCUR; SO TO PRESERVE ITS GENETIC INFORMATION, THE ORIGINAL BRAMLEY IS BEING CLONED.

Bess Pool NOTTINGHAMSHIRE
Keswick Codlin LANCASHIRE
Tower of Glamis SCOTLAND
Lord Burghley LINCONSHIRE
Devonshire Quarrenden
George Neal KENT
Claygate Pearmain SURREY
Tom Putt SOMERSET
Brownlees Russet HERTFORDSHIRE
Peasegood Nonsuch LINCONSHIRE

During the wassail, a toast is drunk to the apple tree to bid it good health for the forthcoming season, as shown in this engraving from the Illustrated London News, 1861.

The origins of wassailing are lost in the mists of time, but the word comes from 'wes hal', an old English salutation meaning 'Be thou of good health'. It took place any time during the twelve days of Christmas, but the most usual date was Twelfth Night (6 January) or Old Twelfth Night (17 January by the old Julian calendar).

The ceremony was held to ensure the well-being of the trees and thus a plentiful apple crop. The order and ritual of the ceremony varied in different parts of the country but always involved making a lot of noise to frighten away any evil spirits living in the apple trees. The owner, his workmen, and their families would gather in the orchard, fire shotguns through the branches and bang pots and pans together. Cider would be poured around the tree roots to sustain it in the forthcoming year, and pieces of toast or small cakes soaked in cider placed in the branches to feed the tree spirits, which were sometimes represented by the robin. In some places, the branches of the trees were pulled down and the buds dipped in cider. In another variation, the men would bow down to the ground three times and rise up slowly as if lifting a heavy sack of apples, to show the trees what was required. The wassail song was sung throughout, asking the trees to bear good crops: 'Old Apple tree, we wassail thee/And hoping thou wilt bear/Hatfuls, capfuls, three bushel bagfuls/And a little heap under the stairs./Hip! Hip! Hooray!'

Wassailing was an essential part of the apple tree's year; it was considered very unlucky if the ceremony did not take place. The practice had largely died out by the early twentieth century but recently it has undergone something of a revival in traditional orchard-growing areas. In Devon, Buckland Abbey and Killerton hold wassail ceremonies on Old Twelfth Night to celebrate the custom and have some fun at the same time.

The Wassail Bowl

In some places, a wassail bowl was taken from house to house during the Christmas period: householders would exchange money and good wishes for a taste of the drink. The bowl was usually made from apple wood and the brew it contained was a potent mixture of ale, sugar, spices and roasted apples. For an updated version, bake three small red apples in the oven, basting with sugar and a small amount of brown ale until soft. Heat 2 pints (1 litre) of brown ale with a 1/2 pint (1/4 litre)of sweet sherry and a large pinch each of cinnamon, nutmeg and ginger and a strip of lemon rind. Simmer for five minutes, then add the baked apples and brown sugar to taste.

Thomas Knight (1758-1838) was the first to experiment with cross-pollination of apple blossom using named varieties. Before then, cross-pollination was a matter of serendipity, with insects carrying any mixture of pollen from blossom to blossom, giving each apple when formed the potential of growing a unique variety from each of its pips.

'For what is more lovely than the bloom of orchard trees in April and May, with the grass below in its strong, young growth; in itself a garden of cowslips and daffodils?' The words of the eminent gardener Gertrude Jekyll are as true today as when she wrote her book *Colour in the Flower Garden* over ninety years ago. A traditional orchard in bloom is the quintessential feature of the English landscape; well-grown trees laden with pink and white blossom, the air redolent with its delicate scent and full of the sound of birdsong and bees, whilst beneath the trees, sheep graze amongst wild flowers.

The key to a good crop of apples is the successful pollination of the blossom. This is not as straightforward as it sounds; most apples set a better crop if pollinated by another variety which flowers at the same time. For the purpose of pollination, apples are grouped together according to the time that they flower, with the earliest flowers in group 1 and the latest in group 7. Amongst the first apple variety to bloom is the American variety Scarlet Pimpernel (also known as Stark's Earliest), whilst the Nottinghamshire apple Bess Pool (group 6) is one of the latest to flower and thus ideal for planting in areas where there is a danger of late frosts. For a good crop, Bess Pool should be planted with another late flowerer from the same group, such as Court Pendu Plat or with a variety where the blossoming period overlaps, like Norfolk Royal in group 5.

The majority of apples fall into pollination groups 2, 3 and 4 so there are plenty of varieties to choose from with compatible pollination periods. A few varieties, such as Bramley's Seedling and Ribston Pippin, do not produce good pollen and cannot reciprocate in the pollination process, so two other pollinators should be planted. When space is at a premium and there is only room for one apple tree, plant a family tree, which has more than one compatible variety grafted on to the rootstock. Crab apples can also be useful pollinators.

Insects such as bees are an essential part of this cross pollination process and should be encouraged. An orchard is the ideal spot for a bee hive; bees and butterflies will also be attracted by other scented flowers. Apple trees are prey to a number of pests and diseases; good husbandry and hygiene can help to prevent some of these, but if pesticides are employed, they should never be used at blossom time to avoid harming the beneficial pollinating insects.

The apple tree Golden Reinette at Westbury Court Garden, Gloucestershire, where it is grown as an espalier. The branches are trained horizontally and carefully pruned to maintain their shape.

If you want the best for your apple trees, there are many points to consider beyond the complexities of pollination described on page 36. Here are a few tips from National Trust gardeners:

- Choose local varieties: apples are the most adaptable of trees and there are few areas in Britain where they cannot be grown. But unless you have lots of time to spare, it makes sense to plant a variety that has originated in your area and has therefore evolved to suit the local climate and soil conditions. The damper conditions of the West Country, for example, are not ideal for varieties like Cox's Orange Pippin which are prone to scab and other fungal diseases prevalent in wet areas. Always check with the nursery that the tree of your choice is growing on a rootstock suitable for your garden; M25 rootstock will produce a vigorous tree of up to 20ft high in a few years, whereas M9 rootstock will result in a tree of 6-8ft.

- Less pruning means fewer pests and diseases advises Barry Champion of Trelissick. Whilst young trees need formative pruning in order to achieve the shape you require, after that it should largely be confined to removing any diseased or crossing branches and done in winter. The general aim should be to open up the tree to allow in light and air. If it is an old tree which has not been touched for a while, prune it over four years or so, rather than in one go. Trained trees, such as fans and espaliers, require more pruning to keep them in shape; this is usually done in early August. Disinfect your secateurs between each tree so that diseases don't get passed on.

- Of all the pests which can damage an apple crop, the codling moth is the worst. Eggs are laid on the fruit by wingless females and the caterpillars eat into the centre of the apple. Tie grease bands (or strips of sheepskin with the wool side out) around the tree trunk to prevent the females climbing up; caterpillars will pupate in corrugated card hung around the branches which can then be collected and burnt.

- Get a good book on the care of fruit trees (see page 48); don't get too hung up on all the technical detail and above all enjoy your apples!

Good companions

Folk lore: nasturtiums planted around the base of apple trees will prevent woolly aphids from infecting the trees. Chives are also good companions, whereas apples and carrots should not be planted together if you want your plants to stay healthy and bear well.

APPLE
REINETTE
MID-17th CENTURY

An idyllic scene of apple harvesting as seen through the eyes of the Victorians, from the Illustrated London News, *1873.*

The apple harvest begins in August when late summer dessert apples such as Beauty of Bath, Devonshire Quarrenden and Lady Sudeley are ready for picking. These first apples will not keep and should be eaten straight from the tree or within the week. An apple picked in its prime is a different fruit from one harvested too early so it is worth taking the trouble to pick at the right time. The fruit is ripe when it parts easily from the twig; hold the apple in the palm of your hand and twist slightly. Windfalls on the ground and the pips turning from white to brown are other signs that the fruit is ripe. Even then, ripening times of fruit on one tree will differ according to how much sun each apple gets.

The latest varieties to ripen are the spring dessert apples which should be picked as late as possible in October (any later and they are likely to be blown down or damaged by rain) and then allowed to ripen in store when they will keep until March or April. These useful apples include such varieties as Cornish Gilliflower, D'Arcy Spice and Braddick's Nonpareil. At Felbrigg, Head Gardener Ted Bullock keeps his Norfolk Beefings until the spring, by which time this culinary apple is sweet enough to eat as a dessert.

As with most areas of apple cultivation, the harvest was surrounded in superstition. In some parts of the country, it was believed that the apples should not be picked when the moon was waxing, otherwise they would not keep through the winter. Conversely in areas of the West Country, it was unlucky to harvest the fruit when the moon was on the wane. It was common practice to leave a few apples on the tree after the bulk had been picked to feed the guardian spirits of the trees but it was bad luck if the fruit was still there in the spring. Whether you believe in feeding the spirits or not, the decorative effect of a few late apples left on the tree during the winter months can lift the spirits on a chill January day.

Unlike the heap of apples shown in this picture, apples must be kept separately in storage, otherwise varieties such as Colonel Vaughan and St Edmund's Pippin which are late autumn desserts will hasten the ripening of later ones like Easter Orange and Hoary Morning.

In the commercial fruit industry, when apples are taken into store the atmosphere is controlled to inhibit the maturing of the apples and keep them fresh for longer. Thus Bramley's Seedling is available throughout the year and the life of a Cox's Orange Pippin can be extended to eight months from the time it was picked. Once commercially stored apples are past their best, there are plenty of imports from New Zealand and South Africa on offer. But the taste of these imports does not always live up to their attractive appearance; they are usually picked before they are ready to eat and then ripened in transit. There is nothing to beat the satisfaction (and taste) of eating your own and by choosing the right varieties and careful storage it is possible to eat home grown apples from August until April or even May.

Apples were once stored in straw in the ground, but the more usual practice is to store them on racks in single layers in a cool, dark frost-free shed. Scattering water on the floor from time to time will help to maintain the humidity and prolong the life of the apples. Only keep unblemished fruit, and store mid-season apples separately from late apples as the ethylene given off by the early fruit will hasten the ripening of the later varieties. Opinions differ as to whether the apples should be wrapped in paper and much depends on time available and the numbers of apples stored; although this practice can stop the spread of rot, it is more time consuming to unwrap each one to check if you are less than diligent. Another method is to store the apples in clear polythene bags with holes for ventilation.

There is of course more than one way to store an apple, especially if it is not a good keeper. In her book *Food in England*, Dorothy Hartley recommends threading peeled and cored apples on a string to make a long necklace which is then hung in a dry airy place or warm room to dry. Another method is to thread the apple rings on sticks which are placed across drying trays and put near the fire or in a cool oven. After five or six hours the apples should be ready with a texture resembling chamois leather. Once cooled and aired, the dried apple rings can be bottled in dry jars.

LAXTONS EPICURE
PITMASTON PINEAPPLE
ST. EDMUNDS PIPPIN
RIBSTON PIPPIN
ADAMS PIPPIN
EASTER ORANGE
SPARTAN
ROSS NONPAREIL
PITMASTON PINEAPPLE
KESWICK CODLIN
RIBSTON PIPPIN
INGRID MARIE
KING GEORGE V
ADAMS PEARMAIN
EGREMONT RUSSETT
MERTON CHARM
MERTON KNAVE
ADAMS PEARMAIN
CLAYGATE PEARMAIN
WINSTON
DOMINO
DEVONSHIRE QUARRENDEN
GEORGE NEAL
DR. HARVEY
LAXTONS REWARD
SPARTAN
PEASGOODS NONESUCH
STAR of DEVON
ARTHUR TURNER
WEALTHY
TYDEMANS EARLY
JOHN DOWNEY
FORTUNE
GOLDEN HORNET
HOARY MORNING

The pips, pulp and peel of the apple were used as a means of divination. In this engraving by Winslow Homer, young people of both sexes are shown peeling apples to take part in a game where the unbroken peel was thrown over the shoulder to reveal the initial of their future partner.

As befits something so rooted in history, the apple is the stuff of legends. References to the fruit abound in classical mythology and Celtic stories, where they are usually associated with love or immortality. It was an apple that Paris gave to Aphrodite to show she was the most beautiful of the goddesses. Aphrodite gave apples to help Hippomenes win the hand of Atalanta, a proud and athletic runner who would only marry the man who could outrace her. The apples were irresistible to women and whenever Atalanta was ahead in the race, Hippomenes would throw down an apple to distract her and thus won the race and Atalanta. In Celtic legend, King Arthur was taken to the Isle of Avalon – an island paradise of apple trees – to be healed from his fatal wound and it was here that good knights went after death before they could be restored to the mortal world.

The apple as a means of divination is probably derived from the Druidic religion, in which apples were held in high esteem as the host plant for the sacred mistletoe. Any number of customs relating to foretelling the future are associated with the apple and, again, usually involve love. In one game, the apple is peeled in one long strip and the peel is thrown over the left shoulder to form the initial of one's future partner. At Halloween, a girl would put an apple under her pillow to dream of the man she would marry, whilst young singles of both sexes would find out which of them would be the first to marry by each twirling an apple on a string; the owner of the apple which fell off first would be the lucky one. Games such as apple bobbing and apple on a string – where players have to try to bite an apple suspended on a string without using their hands – are still played today and new ones have been invented as part of Apple Day celebrations.

Personalised Apples

Fun for children: cut a small pattern or your initials out of paper and paste to the side of the apple exposed to the sun just before it starts to turn red. When the fruit is ripe, remove the paper and the pattern is repeated on the skin of the apple.

W HOMER - DEL

Useful Addresses

National Trust Properties

There are Trust properties throughout the country with orchards and walled gardens which feature collections of old varieties of apples, some of which are mentioned in the text:

Acorn Bank, Temple Sowerby, nr Penwith, Cumbria CA10 1SP
Beningbrough Hall, Beningbrough, York YO30 1DD
Berrington Hall, nr Leominster, Herefordshire HR6 0DW
Barrington Court, Barrington, nr Ilminster, Somerset TA19 0NQ
Coughton Court, nr Alcester, Warwickshire BA94 5JA
Cotehele, St Dominick, nr Saltash, Cornwall PL12 6TA
Erddig, nr Wrexham, North Wales LL13 0YT
Felbrigg Hall, Felbrigg, Norwich NR11 8PR
Killerton, Broadclyst, Exeter, Devon EX5 3LE
Nunnington Hall, Nunnington, York YO62 5UY
Quarry Bank Mill, Wilmslow, Cheshire SK9 4LA
Trelissick, Feock, nr Truro, Cornwall TR3 6QL
Trerice, nr Newquay, Cornwall TR8 4PG
Westbury Court, Westbury-on-Severn, Gloucestershire GL14 1PD

You can visit these properties free if you are member of the National Trust and help it in its conservation work. Ask for details at any of the properties listed above or contact the National Trust, PO Box 39, Bromley, Kent BR1 3XL (0181 315 1111) e-mail: enquiries@ntrust.org.uk

National Fruit Collection, Brogdale Horticultural Trust,
Brogdale Road, Faversham, Kent ME13 8XZ (01795 535 286)
See page 8. You can also join the Friends of Brogdale which was set up to support the Horticultural Trust and encourage appreciation of apples and other fruit. Membership entitles you to free entry to Brogdale and regular information on events such as orchard tours, sales of fruit and trees, demonstrations and fruit identification.

Royal Horticultural Society, Wisley Gardens, Surrey GU23 6QB (01483 224 234)
Holds a collection of over five hundred apple varieties; fruit trees and fruit for sale and fruit identification. There is also an RHS Fruit Group based at the RHS, 80 Vincent Square, London SWAP 2PE with an annual newsletter and events held in the Home Counties, Midlands, East Anglia and the West Country.

Priorwood Gardens, Melrose, Roxburghshire, Scotland (01896 822 965)
An orchard of apples from Roman times to the present day, including local varieties, looked after by the National Trust for Scotland.

Common Ground, PO Box 25309, London NW5 1ZA (0171 267 2144)
See page 8. Publications and information about apples, including a list of Apple Day events and local apple groups.

Old varieties of apples may be rediscovered at *Apple Day* events like this one at Brogdale.

Recommended Reading

This book can only hope to act as a taster to the fascinating history of apples. If, like Eve, you are tempted to find out more, the following books – some of which have been consulted during the research for this book – are recommended:

Harry Baker, *The Fruit Garden Displayed*, Royal Horticultural Society, 1986
A practical guide on the cultivation and care of fruit.

Francesca Greenoak, *Fruit and Vegetable Gardens*, The National Trust/Pavilion, 1990
A practical guide to some of the National Trust's orchards, potagers and kitchen gardens. Currently out of print.

Joan Morgan and Alison Richards, *The Book of Apples*, Ebury Press, 1993
The definitive book for everything you wanted to know about apples, with a detailed history and practical advice, an apple directory of over 2,000 varieties and tasting notes.

Sara Paston-Williams, *Jams, Preserves and Edible Gifts*, The National Trust, 1999

Sara Paston-Williams, *The National Trust Book of Christmas and Festive Day Recipes*, Penguin Books, 1983
Currently out of print.

Rosanne Sanders, *The English Apple*, Phaidon Press, 1988
Detailed descriptions and wonderful paintings of 122 varieties of apple and blossom. Currently out of print.

Muriel W.G. Smith, *The National Apple Register of the United Kingdom*, Ministry of Agriculture, Fisheries and Food, 1971
Much used by apple collectors with its listing of over 6,000 apple names and details of sources of information.

Virginia Spiers, *Burcombes, Queenies and Colloggetts*, West Brendon, 1996
An inspirational book about the work of the artist Mary Martin and her partner James Evans in saving near extinct varieties of apples, pears and cherries in the Tamar Valley in Cornwall, illustrated by Mary's paintings of fruit and orchards.

Harold Taylor, *The Apples of England*, Crosby Lockwood and Sons, 1945

Ruth Ward, *A Harvest of Apples*, Sage Press, 1997
History, anecdotes and apple recipes from drinks to desserts, preserves and pies.

Various, *Orchards, A Guide to Local Conservation*, Common Ground, 1989

Various, *Apple Games and Customs*, Common Ground, 1994